Buildi
Environments With
Retaining Walls

Schiffer
Publishing Ltd®
4880 Lower Valley Road, Atglen, PA 19310 USA

Published by Schiffer Publishing Ltd.
4880 Lower Valley Road
Atglen, PA 19310
Phone: (610) 593-1777; Fax: (610) 593-2002
E-mail: Info@schifferbooks.com

For the largest selection of fine reference books on this and related subjects, please visit our web site at **www.schifferbooks.com**
We are always looking for people to write books on new and related subjects. If you have an idea for a book please contact us at the above address.

This book may be purchased from the publisher.
Include $3.95 for shipping.
Please try your bookstore first.
You may write for a free catalog.

In Europe, Schiffer books are distributed by
Bushwood Books
6 Marksbury Ave.
Kew Gardens
Surrey TW9 4JF England
Phone: 44 (0) 20 8392-8585;
Fax: 44 (0) 20 8392-9876
E-mail: info@bushwoodbooks.co.uk
Website: www.bushwoodbooks.co.uk
Free postage in the U.K., Europe; air mail at cost.

Copyright © 2006 by Schiffer Publishing, Ltd.
Library of Congress Control Number: 2006928947

Type set in Geometric 231 Hv BT/Zurich BT

ISBN: 0-7643-2542-6
Printed in China

Foreword

Since their introduction in the 1980s, segmental retaining walls (SRWs) have revolutionized the landscaping industry. Less costly and infinitely more attractive than poured-in-place concrete, and virtually maintenance-free, SRWs opened up a whole new world of design for landscape architects. Long-lasting durable concrete offers an environmentally sound alternative to treated landscape timbers or creosote railroad ties. But perhaps best of all, SRWs made it possible for amateur landscapers and do-it-yourselfers to turn their own back yards into stunning outdoor environments.

The possibilities are limitless with the amazing flexibility of today's segmental retaining walls. But these systems aren't just for retaining walls anymore. Versa-Lok's uniquely engineered design, introduced as the original solid, pinned system in 1987, enables the creation of variable-radius curves, multiangle corners, stairs, columns and freestanding walls. Creative landscape architects also are finding ways to use Versa-Lok in building outdoor kitchens, spas and entertainment areas. Available in a variety of regional colors and two distinctive textures—classic split-face and vintage Weathered—Versa-Lok retaining wall systems provide landscape professionals with a full palette of creative options.

In addition to a complete how-to section for installing your own retaining walls, Building Outdoor Environments with Retaining Walls features a wealth of images that beautifully illustrate what's possible with Versa-Lok. If you're a homeowner looking for ideas to transform your yard, look no further. If you're a landscape architect, designer or contractor, there's plenty here to inspire you as well.

Let Building Outdoor Environments with Retaining Walls be your guide to the 21st century solution preferred by landscape architects, designers and do-it-yourselfers everywhere.

Todd Strand
President
Versa-Lok Retaining Wall Systems
Oakdale, Minnesota

Contents

Acknowledgments

The following people contributed to this book:

Editor in Chief: Tina Skinner
Art and Copy Director: Karl Bremer
Editorial Advisor: Jeffrey Snyder
Text Coordinator: Lindsey Hamilton
Electronic Production Coordinator: Cathy Buck
Book Design: John Cheek
Cover Design: Bruce Waters

VERSA-LOK has been creating "Solid Solutions™" for architects, engineers, contractors, and homeowners since 1987. VERSA-LOK systems are the original solid, pinned segmental retaining walls (SRWs). Quick and easy to install, VERSA-LOK retaining walls offer superior durability and a wealth of design/build options, along with a variety of styles and sizes. Whatever your landscape needs, VERSA-LOK has a retaining wall for you. For more information, visit www.versa-lok.com or call 1-800-770-4525.

Introduction

This book is a collection of projects and an installation guide by Versa-Lok® Retaining Wall Systems. The craftsmanship and expertise will hopefully inspire readers to expand their landscapes. From outdoor entertaining areas to luxurious swimming pools, it's amazing what a wall can do. Versa-Lok Retaining Walls present numerous opportunities to transform life in the backyard. Outdoor areas encourage entertaining, relaxation, and creativity.

The installation guide will provide easy to follow instructions and a glossary of terminology to ease the process. It demonstrates every aspect of retaining walls, from the first shovel to the final flower. Consumers will learn the difference between a reinforced and stand-alone wall, why drainage is important and how to cap a wall. The gallery hosts a plethora of projects made with Versa-Lok blocks and know-how. The projects range from backyard patios to forty-foot high walls. The selection makes this book a great reference for contractors, landscapers, and builders. Any potential client will be floored by these walls.

Versa-Lok is based in Oakdale, Minnesota, and has been creating solid, dependable retaining walls since 1987. Much of the information inside was taken

from their installation guides. More literature can be found at their website, www.versa-lok.com. They joined with Schiffer Publishing to produce an informative text to consumers featuring both installation guides and do-it-yourself ideas.

How to Install a Retaining Wall

Installation Guide

With the proper tools and careful planning, installing a segmented retaining wall can be a fun project with beautiful results. Using the VERSA-LOK system of interlocking units and mortarless construction will ensure a durable, attractive, and maintainable retaining wall.

VERSA-LOK Standard retaining wall units are ideal for residential, commercial, and agency projects. They are made from high-strength, low absorption concrete on standard block machines. Solid characteristics make Standard units resistant to damage before, during, and after construction in all climates.

Holes and slots molded into units accept VERSA-TUFF Pins®, which are non-corrosive, glass reinforced nylon pins. Pins interlock units and help provide consistent alignment. This unique hole-to-slot pinning system permits easy variable-bond construction – keeping vertical joints tight.

VERSA-LOK Standard Units

Standard is Versa-Lok's original and most popular unit. Contractors, engineers, and landscape architects agree its solid construction and unique pinning system provide an endless array of design options, hardcore durability and the fastest, easiest installation available. Also available in an increasingly popular Weathered texture at selected dealers.

Standard System Overview

Pinning

VERSA-LOK Standard units interlock with non-corrosive VERSA-TUFF Pins (two per unit). As wall courses are installed, pins are inserted through holes in uppermost course units and are received in slots of adjacent lower course units. Pinning helps to align units in a consistent 3/4-inch setback per course.

Unreinforced Walls

On many projects, VERSA-LOK Standard retaining walls work purely as gravity systems– unit weight alone provides resistance to earth pressures. Frictional forces between units and pin connections hold units together so walls behave as coherent structures. Batter setback of wall faces offers additional resistance against overturning.

Maximum allowable wall height for gravity walls varies with soil and loading conditions. Generally, with level backfill, good soils, and no excessive loading, VERSA-LOK Standard gravity walls are stable to heights of four feet.

Reinforced Walls

When weight of units alone is not enough to resist soil loads, horizontal layers of geosynthetics are used to reinforce soil behind walls. With proper soil reinforcement and design, VERSA-LOK Standard walls can be constructed to heights in excess of 50 feet. Geosynthetics do not act as tiebacks for wall faces. Rather, geosynthetics and soil combine to create reinforced soil structures that are strong and massive

Standard Unit Measurements:	
Height:	6" (152.4mm)
Width (face):	16" (406.4mm)
Depth:	12" (304.8mm)
Face area:	2/3 sq. ft. (.062 sq m)
Weight:	82 lbs. (37.19kg)
Weight/Face Area:	123 lbs./sq. ft (599.84kg/sq m)

enough to resist forces exerted on them. In soil-reinforced walls, Standard units simply retain soil between layers of geosynthetics and provide attractive, durable faces.

Standard Wall Components

This cross section illustrates typical components of VERSA-LOK Standard retaining walls. Mortarless Standard walls are installed on granular leveling pads and do not require concrete footings below frost. The amount and layout of drainage materials and geosynthetic soil reinforcement is site/soil dependent and should be designed by a qualified engineer. The 3/4-inch setback of each unit creates a cant of approximately seven degrees. Canted walls are structurally more stable than vertical walls because gravitational forces "pull" walls into retained soil.

out mortar, they are free to move slightly in relation to each other. Flexibility of the leveling pads and wall units accommodates freeze/thaw cycles without damage to structures. VERSA-LOK Standard walls, installed on granular leveling pads, have been successfully used on projects throughout North America—including shoreline applications and walls exceeding fifty feet in height.

If a contractor chooses to form leveling pads using concrete, unreinforced pads should be made of lean concrete mix (200-300 psi) and no more than two inches thick. To ensure correct Standard unit alignment, special care needs to be taken to construct concrete pads that are exactly level. In rare situations where rigid, reinforced concrete footings are required, they should be placed below seasonal frost depths. Compacted granular leveling pads provide a stiff but flexible base.

Foundation

Foundation soils on which segmental retaining walls will rest must be stiff, firm, and have sufficient capacity to support wall system weight. Any loose, soft, or compressible material must be removed and replaced with properly compacted backfill. The bearing capacity of the foundation soils should be addressed by a soils engineer.

Mortarless VERSA-LOK Standard retaining walls do not require rigid concrete footings below frost.

VERSA-LOK Standard retaining walls are installed on leveling pads consisting of coarse sand or well-graded angular gravel. The most commonly used material for leveling pads is that which is used locally as road base aggregate. Granular leveling pads provide stiff, yet somewhat flexible, bases to distribute wall weights. Rigid concrete footings extending below frost are not required or recommended. Because Standard units are installed with-

Flexibility of the leveling pads and wall units accommodates freeze/thaw cycles without damage to structures.

Embedment

VERSA-LOK Standard segmental retaining walls usually have one-tenth of exposed wall heights embedded below grade or a minimum of one course of block. For example, a wall with ten feet of height exposed above grade would have a minimum of one foot buried below grade – making a total wall height of eleven feet. Embedment should be increased for special conditions such as slope at the toe of walls, soft foundation soils, or shoreline applications. Embedment provides enhanced wall stability and long-term protection for leveling pads.

Deep below the water level, leveling pads and extra embedment measures are working hard to provide a solid area. Elements such as shorelines, surcharge loads, and special soils had to be considered in the design and construction.

Soils and Compaction

With proper design, segmental retaining walls can be constructed within a wide variety of soil conditions. Granular soils are preferred as fill in the areas reinforced with geosynthetics; however, fine-grained soils such as clays are acceptable. Usually, coarse soils require less soil reinforcement and are easier to compact than fine soils. Problem materials like expansive clays, compressible soils, or highly organic soils (top soils) should be avoided or properly addressed in designs.

Proper compaction of foundation and backfill soil is critical to long-term performance of retaining wall systems. Loose backfill will add pressure on walls, collect water, cause settlement, and will not anchor soil reinforcement materials properly. Foundation and backfill materials should be compacted to at least ninety-five percent of standard Proctor density. (Proctor density is the maximum density of the soil achieved in a laboratory using a standard amount of compaction effort.) Gener-

ally, construction, observation, and testing for proper soil type and compaction is provided by the project's soils engineer.

Drainage Within Walls

Segmental retaining walls are designed assuming no hydrostatic pressure behind walls. Drainage aggregate (angular gravel, free of fine soil particles) placed behind walls helps eliminate water accumulation. Because no mortar is used in VERSA-LOK Standard wall construction, water is free to weep through joints of installed units. For walls greater than three feet in height, a perforated drainpipe is recommended at the base of the drainage aggregate to quickly remove large amounts of water.

If high groundwater levels are anticipated or if the wall is along a shoreline, additional drainage materials behind and below reinforced fill may be required. Filter fabric may be required to prevent unwanted migration of fine soil particles into the drainage aggregate.

Surface Drainage

Wall sites should be graded to avoid water flows, concentrations, or pools behind retaining walls. If swales are designed at the top of walls, properly line and slope them so water is removed before it can flow down behind walls.

Give special attention to sources of storm water from building roofs, gutter downspouts, paved areas draining to one point, or valleys in topography. Be sure to guide flows from these areas away from retaining walls. Slope the soil slightly down and away from wall bases to eliminate water running along bases and eroding soil. If finish grading, landscaping, or paving is not completed immediately after wall installation, temporarily protect the wall from water runoff until adjacent construction and drainage control structures are completed.

Water should be directed away from walls with drainage structures.

Geosynthetic Reinforcements

Geosynthetics are durable, high-strength polymer products designed for use as soil reinforcement. Horizontal layers of geosynthetics provide tensile strength to hold the reinforced soil together, so it behaves as one coherent mass. The geosynthetic reinforced soil mass becomes the retaining wall. Sufficient length and strength of geosynthetics can create a reinforced soil mass large enough and strong enough to resist destabilizing loads. Geosynthetic layers also connect the VERSA-LOK Standard units to the reinforced soil.

This lovely lawn could turn into a messy mud pit without a proper water drainage system.

Geosynthetics are made from several types of polymers that resist installation damage and long-term degradation. Geosynthetics are designed to interact with the soil for anchorage against pullout and resistance to sliding. Geogrids, the most common soil reinforcement for walls, are formed with an open, grid-like configuration. Geotextiles (solid fabrics) are also used. Product-specific testing determines the durability, soil interaction, and strength of each type of geosynthetic.

The interaction of various geosynthetics with Versa-Lok units (connection strength) is also thoroughly tested.

Geosynthetic layers must be nominally tensioned and free of wrinkles when placed. Geosynthetics are generally stronger in one direction—the roll direction. It is important that the high-strength direction be placed perpendicular to the wall face, in one continuous sheet (no splices). Along the wall length and parallel to the face, adjacent sec-

tions of reinforcement are placed immediately next to each other without overlap to create 100 percent coverage with no gapping, and with special details for curves and corners.

The required type, length, vertical spacing, and strength of geosynthetic vary with each project depending on wall height, loading, slopes, and soil conditions. A professional Civil Engineer (P.E.) should prepare a final, geogrid-reinforced wall design for each project.

Engineering

VERSA-LOK walls are designed as traditional gravity walls. For unreinforced walls, the stabilizing weight of the battered wall units is compared to the loading on the walls to ensure stability against overturning and sliding. When the loading exceeds the stability of the units alone, a larger gravity mass is created from reinforced soil.

To ensure stability of a reinforced retaining wall, the wall engineer must design the reinforced soil mass large enough to resist loads from outside the wall system (external stability) and with enough layers of proper strength geosynthetic to keep the reinforced soil mass together (internal stability). In addition, the design must have sufficient geosynthetic layers to keep VERSA-LOK Standard units stable and properly connected to the reinforced soil mass (facial stability).

For internal stability, the geosynthetic layers must resist loads that could pull apart the reinforced soil mass. The wall design engineer must ensure the geosynthetic has enough anchorage length to resist pullout from the stable soils and enough strength to resist overstress (breakage). The geosynthetic also must be long enough to resist sliding along the lowest layer.

For external stability, the reinforced soil mass must have sufficient width to resist sliding and overturning. The wall design engineer increases geosynthetic lengths until the reinforced soil is massive enough to provide required stability. The project geotechnical engineer should review the wall design and site soil conditions for external stability against bearing failures, settlement, or slope instability. Often, the wall design engineer can address any such global stability concerns by increasing geosynthetic lengths.

For facial stability, the wall design engineer must ensure wall units can resist loads at the face of the wall, and stay connected to the reinforced soil mass, stay interlocked between geosynthetic layers, and not overturn at the top of the wall.

Loading on segmental walls is dependent on soil conditions, surcharges, slopes, water conditions, and wall heights. Accurate knowledge of each of these properties is needed for a proper design. Soil properties required for a segmental retaining wall design include the internal friction angle and soil unit weight. Generally, the cohesion (c) of any fine-grained soils is conservatively ignored to simplify the design.

Special Design Considerations

Shorelines
VERSA-LOK Standard retaining walls perform well in shoreline applications. However, special design considerations are often necessary to ensure that water pressures do not build up behind walls. Special provisions may include granular reinforced backfill, additional drainage aggregate, drainage behind reinforced soil masses, and filter fabric. Protection of bases from water scour, wave action, and ice may also be necessary. With proper design and reinforcement, VERSA-LOK Standard walls can accommodate special site conditions such as water loads, slopes, or surcharges.

The border of rocks along this river protect the base of this wall while serving as a flood barrier.

Loads Behind Walls
Surcharge loads behind walls can substantially increase amounts of required soil reinforcement. Common surcharge loads include parking areas, driveways, roads, and building structures. For design purposes, permanent loads like buildings are considered to contribute to both destabilizing and stabilizing forces acting on walls. Dynamic forces like vehicular traffic are considered to contribute to destabilizing forces only. Often, the highest surcharge loads are caused by grading or paving equipment during construction. Heavy equipment should be kept at least three feet behind the back of retaining wall units. Soil reinforcement designs should accommodate all anticipated surcharge loads—even if they will occur infrequently or just once.

The weight of cars and water pressure from the pond require extra reinforcements for this parking area to be secure.

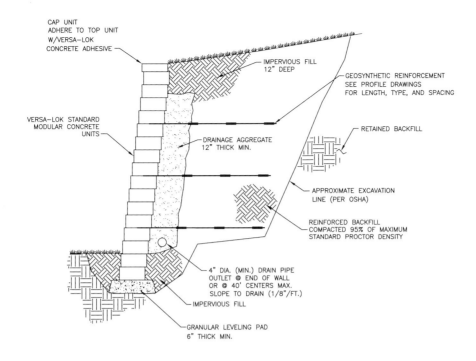

CAP UNIT
ADHERE TO TOP UNIT
W/VERSA-LOK
CONCRETE ADHESIVE

IMPERVIOUS FILL
12" DEEP

GEOSYNTHETIC REINFORCEMENT
SEE PROFILE DRAWINGS
FOR LENGTH, TYPE, AND SPACING

VERSA-LOK STANDARD
MODULAR CONCRETE
UNITS

RETAINED BACKFILL

DRAINAGE AGGREGATE
12" THICK MIN.

APPROXIMATE EXCAVATION
LINE (PER OSHA)

REINFORCED BACKFILL
COMPACTED 95% OF MAXIMUM
STANDARD PROCTOR DENSITY

4" DIA. (MIN.) DRAIN PIPE
OUTLET @ END OF WALL
OR @ 40' CENTERS MAX.
SLOPE TO DRAIN (1/8"/FT.)

IMPERVIOUS FILL

GRANULAR LEVELING PAD
6" THICK MIN.

TYPICAL SECTION—REINFORCED RETAINING WALL
MODULAR CONCRETE UNIT
SCALE: NONE

Tiering

Aesthetically, it may sometimes be desirable to divide large grade changes into tiered wall sections. However, upper wall tiers can add surcharge loads to lower walls and necessitate special designs. To avoid loading lower walls, upper walls must be set back horizontally at least twice the height of the lower walls. If walls are placed closer, lower walls must be designed to resist the load of upper walls. Several closely spaced tiered walls can create steep, unstable slopes. If tiered walls make a grade change steeper than 2:1 (horizontal:vertical), global slope stability may need to be reviewed by a qualified soils engineer.

Slopes

Slopes behind walls increase pressures, sometimes doubling soil loads compared to level backfills. Steep slopes below walls can decrease stability of wall foundations. Slopes can increase the amount of soil reinforcement needed, especially the length. Generally, slopes above or below walls should be no steeper than 2:1 (horizontal:vertical).

Two tiers of wall create a sunken-garden effect in addition to the desired level walkway.

The pressure of this steep hill was divided by two tiered retaining walls. The fences provide an extra safety feature to keep the roadway clear.

Planning, Estimating, and Final Design

Planning

Prior to design, accurate information needs to be gathered including soil conditions, proposed wall heights, topography, groundwater levels, and surface water conditions. Proper permits, owner approvals, utility clearances, and easements should also be obtained.

Make sure that layouts account for minimum curve radii, wall setback, and area needed for geosynthetic soil reinforcement. Be sure that all wall components fit within property constraints. Verify that temporary construction excavations will not undermine foundation supports of any existing structures or utilities. Considerations should also be given to site access for equipment and materials.

Estimating

Accurately estimate and order required materials including VERSA-LOK Standard units, VERSA-TUFF Pins, VERSA-LOK Cap units, VERSA-LOK Concrete Adhesive, imported backfill, leveling pad materials, geosynthetic soil reinforcement, and drainage materials.

For tall walls or complex situations, VERSA-LOK staff engineers can prepare project-specific preliminary designs to be used for estimation purposes.

Final Designs

Final wall designs may be provided prior to putting projects out for bidding. Alternatively, projects can be specified design/build. With design/build projects, the specifiers provide wall layout information (line and grade) but not final engineering for the wall. Contractors submit bids based on this layout including estimated labor, materials, and final engineering costs. Contractors who are awarded projects retain licensed engineers to prepare final wall designs.

A soils report prepared by a qualified geotechnical engineer is needed to provide information on reinforced and retained properties. The soils report should also address slope stability and bearing capacity of foundation soils.

For walls more than four feet in height, most building codes require a final wall design prepared by a licensed Civil Engineer (P. E.) registered in that state. VERSA-LOK and its manufacturers have a network of licensed civil engineers who are familiar with segmental retaining wall design. These individuals are available for referrals to architects, engineers, or contractors with final wall design needs.

Wall Construction

The following tools may be helpful during construction of VERSA-LOK® Standard Retaining Wall Systems.

Tools and Supplies

Safety Protection
Shovel
Four-Foot Level
Smaller Level
Four-Pound Sledge Hammer
Masonry Chisel
Brick Hammer
Tape Measure
Hand Tamper
Vibratory-Plate Compactor
Caulking Gun
String Line
Finishing Trowel
Broom
Diamond-Blade Concrete Saw
Hydraulic Splitter
Transit or Site Level
Backhoe or Skid-Steer Loader
VERSA-LIFTER®

Unit Modification

During wall construction, it will sometimes be necessary to split or cut VERSA-LOK Standard units. Splitting will create attractive, textured surfaces–similar in appearance to front faces of units. Saw-cutting will produce smooth, straight surfaces. In general, units are split when modified portions will be visible. Units are cut when straight edges are required to fit closely next to smooth edges of adjacent units. VERSA-LOK Standard units are easily modified by splitting for a textured face, or by saw-cutting for a smooth side.

VERSA-LOK Standard units are easily modified by splitting for a textured face, or by saw-cutting for a smooth side.

Splitting

To split a VERSA-LOK Standard unit by hand, mark the desired path of split on the unit top, bottom, and back. Score along the top and bottom paths using a two- to three-inch masonry chisel and heavy hammer. Next, place the unit on its face and strike along the back path. It is easier to split units on the ground than on a hard surface. Unit should fracture nicely along paths. If many splits will be required for a project, it may be helpful to rent a mechanical or hydraulic splitter.

Score along the top and bottom paths using a two- to three-inch masonry chisel and heavy hammer. Next, place the unit on its face and strike along the back path.

Saw-Cutting

Saw-cuts are normally made using a gas-powered cut-off saw with a diamond blade. To cut a VERSA-LOK Standard unit, mark the desired path of cut on all unit sides. On a stable work surface, place the unit face toward you with the top side up, at a comfortable height. Make a straight cut down and two to three inches into the face. Move the saw to the top of unit, and cut through top using successively deeper cuts. Flip unit over and finish by cutting completely through the bottom of the unit. If a cut-off saw is not available, a common circular saw and an inexpensive masonry blade may be used. Cut one to two inches deep along the path on the front face. Split the remainder of the unit. The vertical cut on the face of the unit will fit closely against adjacent units – the split portion will not be visible.

Saw-cuts are normally made using a gas-powered cut-off saw with a diamond blade.

Excavation

Carefully plan the location and alignment of the wall base to ensure the top of the wall will be at the desired location.

Excavate just deep enough to accommodate the leveling pad (usually six inches) and required unit embedment below grade. When necessary, also excavate areas where geosynthetic soil reinforcement will be placed. Required unit embedment varies with wall height and site conditions. Generally, if the grade in front of the wall is level, one-tenth of the exposed wall height should be buried (embedded)

below grade, or a minimum of 6 inches. Additional embedment may be required for special conditions including slopes in front of walls, soft foundation soils, and water applications.

Compact soil at the bottom of the excavation. Do not place wall system on loose, soft, wet, or frozen soil—settlement may result. If the wall will sit on previously backfilled excavations such as utility line trenches, be sure the entire depth of existing backfill is well compacted. If necessary, over-excavate soft soils and replace with properly compacted backfill.

Cross-section of an excavated trench. Excavations should be 24 inches wide to allow for Standard units and backfill.

Leveling Pad

Place granular leveling pad material and compact to a smooth, level surface. Leveling pad should be at least six inches thick and twenty-four inches wide. It should consist of coarse-grained sand, gravel, or crushed stone. Use a thin layer of fine sand on top of the leveling pad for final leveling. To quickly construct long sections of leveling pad, create forms by leveling and staking rectangular metal tubing along both sides of the planned pad. Place and compact granular material within these leveled forms and screed off excess.

If the planned grade along the wall front will change elevation, the leveling pad may be stepped in six-inch increments to match the grade change. Always start at the lowest level and work upward. Step the leveling pad often enough to avoid burying extra units while maintaining required unit embedment.

Compact the trench soil to ensure against settlement or unevenness. The leveling pad is an important part of the installation. It should be at least six inches thick and able to support the weight of the retaining wall.

Make sure the leveling pad is completely flat before continuing to the next step. An uneven leveling pad can create major problems.

Base Course

Make sure that the leveling pad is level and begin placing base course units. If the leveling pad is stepped, begin at the lowest point and place the entire length of the lowest course before proceeding to the next course. Align units using their backs or slots rather than their irregularly textured front faces. Stringlines may be helpful when aligning straight walls. Place units side by side on the leveling pad. Front faces of adjacent units should fit tightly and unit bottoms should contact the leveling pad completely. Using a four-foot level, level units front to back, side to side, and with adjacent units. Tap high points with a mallet or hand tamper until level. Take time to ensure a level base course. Minor unevenness in the base course will be amplified and difficult to correct after several courses have been installed.

After the base course has been positioned, place and compact the soil backfill behind the units. Also replace and compact over-excavated soil in front of units at this time. Backfill behind and in front of embedded units should consist of soil—do not use drainage aggregate.

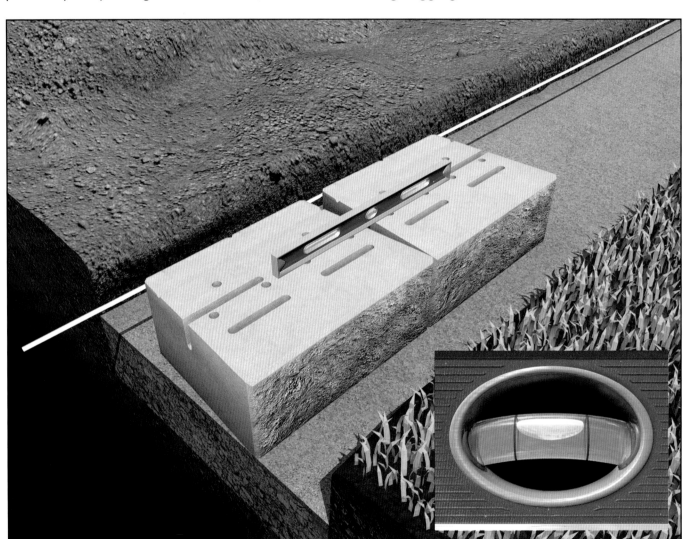

Using a level will guarantee a solid base course.

The base course is the first row of Standard units. This diagram illustrates how to place the units, aligning the front face of the units and leaving a V space in the back to allow for backfill and support. Make sure the units are secure before adding backfill.

Additional Courses

Sweep off tops of installed units to remove any debris that may interfere with additional courses. Place the next course so that the units are set back 3/4 of an inch from the front of the installed units. Set the units a short distance away from their final position and slide them into place. Sliding helps remove imperfections and debris from the top surface of installed units.

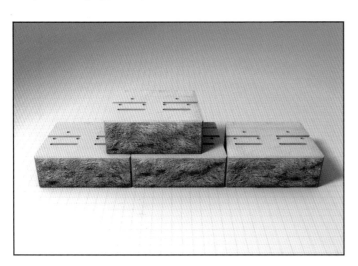

Units should be stacked carefully. The second course should rest a minimum of four inches over the bond of the first course as illustrated.

The VERSA-LOK unique hole-to-slot pinning system allows Standard units to be installed on variable bond. (Units do not need to be placed exactly halfway over the two lower course units.)

The VERSA-TUFF pinning system is a key element in the VERSA-LOK system. Pins are inserted into the outer slots with a mallet.

Vertical joints can wander in relation to other joints throughout walls. However, units should generally overlap adjacent lower course units by at least four inches to aid structural stability. Do no try to install walls on half-bond by leaving gaps in the vertical face; joints can and should be tight.

Insert two VERA-TUFF Pins through the front holes of the upper course units into the receiving slots in the lower course units. There are four front holes in each unit, but only two are used. Use the two outside holes when possible. If one of the outside holes is not usable, move pin to next closest hole. The two pins should engage two separate units in the lower course. Make sure the pins are fully seated in the lower unit slots. If necessary, seat pins using a mallet and another pin. Pins are fully seated when they are recessed approximately one inch below the top surface of upper units.

Pull the units forward to remove any looseness in the pin connection. Check unit alignment and levelness – adjust if necessary. If the length of a course must fit into a limited space or if vertical joints begin to line up with joints in the course immediately below, adjust by installing partial units. Create partial units by saw-cutting whole units into pieces at least four inches wide at the front face. When installing partial units, try to disperse them throughout the wall. This technique helps to hide partial units and lends to a more attractive project.

Stack no more than three courses before backfilling. If VERSA-LOK Standard units are stacked too high, they may push out of alignment during placement of backfill.

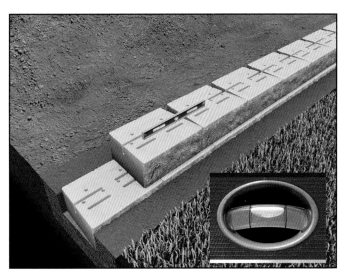

When the bubble stays between the lines on the level, the course of blocks is level and properly installed

Drainage materials

Beginning at the level of the planned grade in front of the wall, place drainage aggregate (3/4-inch clear, free-draining, angular gravel) between and directly behind units to a minimum thickness of 12 inches. Drainage aggregate must be free of fine dirt or soil. Do not place drainage aggregate behind units that will be embedded. Drainage aggregate is critical to wall performance because it keeps water pressures from building up behind the wall face.

For walls over three feet high, perforated drain pipes should be used to collect water along the base of the drainage aggregate. Drain pipes help to quickly remove large amounts of water.

For some projects, often shoreline applications, a geosynthetic filter fabric may be required behind the drainage aggregate. Filter fabric will prevent soils or sands (fines) from migrating into the drainage aggregate and wall face joints.

Proper installation of the drainage aggregate is key to a durable and safe retaining wall. The illustration demonstrates what your project should look like at this point.

Compacted Soil Backfill

Proper placement and compaction of backfill is critical to the stability of a segmental wall. Poorly compacted backfill puts extra pressures on a wall—especially when it becomes wet.

Place soil backfill beginning directly behind drainage fill in layers (lifts) no thicker than six inches. Compact soil backfill—making sure that backfill is neither too wet nor dry. The amount and type of effort needed for adequate backfill compaction varies with soil type and moisture content. Generally, hand-operated vibratory-plate compactors can be used to achieve adequate compaction of granular soils—even on big projects.

Fine soils such as clays should be compacted with kneading-type equipment like sheepsfoot rollers. To avoid pushing wall units out of alignment, do not use heavy self-propelled compaction equipment within three feet of the wall face. At the end of the day's construction, protect the wall and the reinforced backfill from possible rainstorm water damage. Grade the soil backfill so water will run away from wall face and direct runoff from adjacent areas away from the project site.

Compacting backfill against the drainage aggregate will fill out the trench, adding stability and support. Again, a flat tight surface is required.

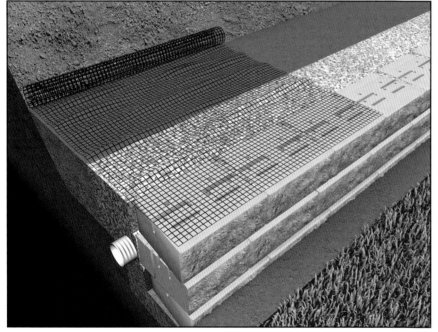

The geogrid provides the majority of the support. Geosynthetics cause the backfill and aggregate to unite as a solid mass. Roll the geosyntheics away from the face of the wall for ultimate strength.

Geosynthetic Soil Reinforcement

Geosynthetic soil reinforcement such as VERSA-Grid® is used to reinforce soil backfill when weight of Standard units alone is not enough to resist soil pressures. Soil reinforcement type, length, and vertical spacing will vary for each project and should be specified in a final wall design prepared by a licensed Civil Engineer (P.E.).

Prepare to install soil reinforcement materials by placing Standard units and backfilling up to the height of the first soil reinforcement layer specified on construction drawings. Lay soil reinforcement horizontally on top of compacted backfill and the Standard units. Geosynthetic layers should be placed about one inch from the front of the Standard units.

Geosynthetics are usually stronger in one direction. It is very important to place them in the correct direction. The strongest direction of the geosynthetic must be perpendicular to the wall face. For correct orientation, follow the geosynthetic manufacturer's directions carefully.

After positioning soil reinforcement, place the next course of Standard units on top of soil reinforcement. Insert pins through Standard units and into lower course units. Place drainage aggregate against the back of the units and on top of the soil reinforcement. Remove slack by pulling soil reinforcement away from the wall face and anchoring at back ends. Beginning at the wall face, place and compact soil backfill. Keep soil reinforcement taut and avoid wrinkles. Place a minimum of six inches of soil backfill before using any tracked equipment on top of soil reinforcement. Follow manufacturer's construction guidelines to avoid damage to soil reinforcement.

Placing soil reinforcement behind curves and corners requires special layout and overlapping procedures. Never overlap soil reinforcement layers directly on top of each other. Slick surfaces of geosynthetics will not hold in place properly when placed directly on top of one another. Always provide at least three inches of soil fill between overlapping soil reinforcement layers.

While spacing of the geogrid will vary, to ensure stability during construction, vertical spacing between geosynthetic layers should never exceed two feet.

More, More, More

Continue placing additional courses, drainage material, compacted soil backfill, and geosynthetic soil reinforcement as specified until desired wall height is achieved. For walls more than four feet high, most building codes require a final wall design prepared by a licensed Civil Engineer (P. E.) registered in that state. VERSA-LOK and its manufacturers have a network of licensed civil engineers who are familiar with segmental retaining wall design. These individuals are available for referrals to architects, engineers, or contractors with final wall design needs.

Verify that the geosynthetic layer is tight and securely fastened with stakes.

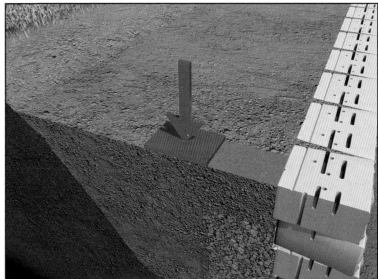

Repeat the backfill, aggregate, and geosyntheic layering, always using levels and compacters.

Once the wall has reached the desired length and height, top it off with Versa-Lok caps.

Overhanging caps slightly on the face of the wall creates an "eyebrow" effect that helps hide minor imperfections in the wall's alignment.

Caps

Finish the wall by placing cap units along the top. Two VERSA-LOK Cap unit types are available—A and B. Alternate A and B caps on straight walls. Use A caps for convex (outside) curves. Use B caps for concave (inside) curves.

If cap layout does not exactly match the wall radius, adjust spacing at the back of the caps —do not gap caps at the front. To completely eliminate gapping, it may be necessary to saw-cut sides of cap units.

Front faces of caps may be placed flush, set back, or slightly hung over faces of VERSA-LOK Standard wall units. It is preferred to overhang cap units approximately 3/4-inch to create an "eyebrow" on top of the wall. Overhanging cap units will create a small shadow on wall units and help to hide minor imperfections in wall alignment.

All cap units should be arranged before securing with VERSA-LOK Concrete Adhesive. Secure caps by placing two, continuous, 1/4-inch beads of adhesive along the top course of wall units. Set caps on prepared wall units. Do not secure caps using mortar or adhesives that become rigid. A VERSA-LOK wall may move slightly (especially in areas subject to freeze/thaw cycles) causing a rigid cap adhesive to fail. Do not place caps if the units are too wet for the adhesive to stick. In cold weather, keep the adhesive tubes warm until just prior to use.

Allow at least two to three days (in warm weather) for adhesive to cure. Thanks to the unique secondary-cure mechanism of VERSA-LOK Concrete Adhesive, its adhesion properties improve with time and weathering. In cold weather, keep the tubes of adhesive warm before applying.

Curves

The trapezoidal shape of VERSA-LOK Standard units permits construction of concave, convex, and serpentine curves. General construction requirements described earlier (leveling pad preparation, drainage, compaction) remain the same for curve installation. All radii distances below are measured from circle centers to front of unit faces.

Concave curves are constructed by increasing spaces between backs of adjacent units—always keeping front joints tightly aligned. Concave curves may be built at any radius; however, a minimum radius of six feet is recommended. Radii smaller than six feet are structurally adequate but tend to appear choppy. Often, it is more appropriate to build inside corners instead of tight concave curves.

Convex curves are constructed by decreasing spaces between backs of adjacent units. Because upper courses of VERSA-LOK Standard units are set back from lower courses by several inches, course radii become smaller as walls become taller. If a course radius becomes too small, Standard units cannot be properly positioned without cutting unit sides. Therefore, careful base course planning for convex curves is important when building tight curves.

Minimum top course radius for convex curves is eight feet. To calculate correct base course radius, add 3/4 inch for each wall course to the minimum radius. For example, minimum base course radius for a wall that will have six setbacks (including embedded units) will be (6 x 3/4") + 8' = 8' 4-1/2".

Corners

Standard units may be easily used to create an unlimited variety of corners. Outside 90-degree corner units are easily created by splitting Standard units in half. Alternate half units as shown on the right. This creates about a

The decorative curves of this wall were created by increasing and decreasing the spaces between the backs of Standard units.

four-inch overlap of the units below. This is acceptable – Standard units do not need to be exactly halfway over the lower units (half-bond).

Half units on outside 90-degree corners do not pin. It is recommended to secure them using VERSA-LOK Concrete Adhesive. No unit modification is necessary to install inside 90-degree corners. Place full-size Standard units as shown, adjusting for proper vertical joint arrangement.

Multiangle corners tamed this steep lot to maximize yard space.

Basic Wall Design Elements

The figures above illustrate a variety of inside and outside corner arrangements. Use these illustrations as guides when designing unique corners. Note that illustrations represent alternating courses and that VERSA-LOK Standard units are modified to create corner units. Split units where textured faces are desired and visible. Saw-cut units when straight edges are needed to fit closely next to adjacent units. Alternating corner units should overlap – do not butt or miter corners. If corners are butted or mitered, differential movement between "separate walls" can occur.

Stepped Base Elevations

If the final grade along the front of the wall changes elevation, the leveling pad and base course may be stepped in six-inch increments to match the grade change. Always start at the lowest level and work upward. Step the leveling pad often enough to avoid burying extra Standard units while maintaining required unit embedment.

Stepped Wall Tops

Wall tops should step to match grade changes. As a wall steps down, use split half units to end each course. Split units provide textured sides to match the wall face. When capping tops of stepped walls, split the exposed side of the last cap unit to create an attractive end.

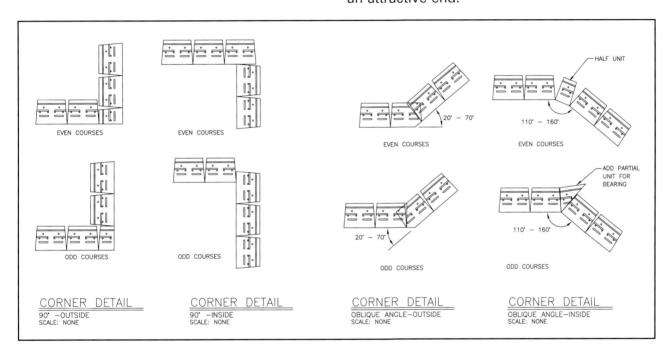

CORNER DETAIL
90° –OUTSIDE
SCALE: NONE

CORNER DETAIL
90° –INSIDE
SCALE: NONE

CORNER DETAIL
OBLIQUE ANGLE–OUTSIDE
SCALE: NONE

CORNER DETAIL
OBLIQUE ANGLE–INSIDE
SCALE: NONE

A retaining wall is staggered to correspond to descending stairs, flaring out in invitation toward the garden.

This stepped wall adds intrigue and greenery to this walkway. The wall increases and decreases in accordance with the grade.

Returns

As an option to stepping wall tops, grade changes at the top of a wall can be accommodated by creating returns that turn into slopes behind a wall. Returns create a terraced appearance instead of several small steps along the top of a wall. The top of VERSA-LOK Standard walls can step down in six-inch increments or in larger steps created by returns.

Advanced Designs

Stairs

Steps with a ratio 2:1 (horizontal:vertical) can be easily installed using VERSA-LOK Standard units. Recommended step construction begins by stacking a pedestal of Standard units. Cap units are then placed as treads and vertical sidewalls are installed.

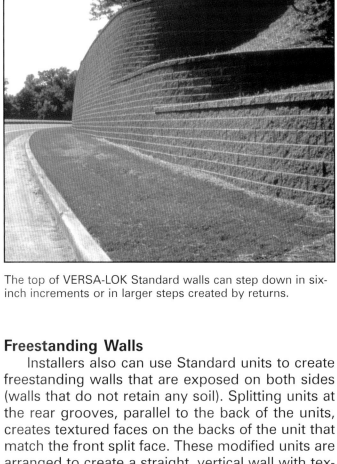

The top of VERSA-LOK Standard walls can step down in six-inch increments or in larger steps created by returns.

This stately staircase was built using a simple ratio of 2:1 and the Standard cap units.

Freestanding Walls

Installers also can use Standard units to create freestanding walls that are exposed on both sides (walls that do not retain any soil). Splitting units at the rear grooves, parallel to the back of the units, creates textured faces on the backs of the unit that match the front split face. These modified units are arranged to create a straight, vertical wall with textured faces on both of the exposed sides of the wall. For stability, freestanding walls should not exceed three feet high.

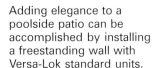

Adding elegance to a poolside patio can be accomplished by installing a freestanding wall with Versa-Lok standard units.

From these split units, installers can build free-standing walls with textured faces on both sides. Designers can use VERSA-LOK freestanding walls for stand-alone walls set directly at grade or for parapets, extending above the top of retaining walls. Similar to VERSA-LOK retaining walls, freestanding wall units interconnect with pins and rest on granular leveling pads. No mortar or concrete footings are needed. The weight of freestanding units and the pinned unit connection provide wall stability. Freestanding walls used as parapets at the top of retaining walls are stable up to 2.5 feet. While VERSA-LOK freestanding walls provide excellent aesthetic options and visual screening, do not rely on them or use them to resist loads such as pedestrian or vehicular traffic. To protect against lateral loads, engineer-designed structures (like guardrails or concrete traffic barriers) should be installed behind walls.

Columns

A wide variety of attractive columns can be easily created from VERSA-LOK Standard units. Columns less than four feet high can be supported on granular leveling pads with no frost footings, just like VERSA-LOK Standard retaining walls. The simplest column is created by splitting Standard units into half-units and vertically stacking them in a 20-inch by 20-inch square column. However, columns of other sizes are also possible with unit modification.

These double columns not only add curb appeal to the home, they require no frost footings and are installed on granular leveling pads.

A freestanding wall with bold pillars complements the landscape.

For stability, taller columns require cast-in-place concrete footings. The center hole of the columns (behind the units) can be used to install steel-reinforced concrete to stabilize taller columns. A qualified professional Civil Engineer should provide a design for columns over four feet high.

Even wildlife can't help but notice the stately elegance of this pillar.

Guide Rails, Railings, and Traffic Barriers

For safety purposes, a variety of barriers may be placed behind VERSA-LOK Standard walls, including fences, railings, and guide rails. Barriers should be placed several feet behind wall faces to provide post foundations. Posts may penetrate geosynthetic soil reinforcement layers in accordance with the manufacturer's and engineer's recommendations.

When space is limited, properly designed, reinforced concrete barriers can be placed directly on top of walls. Expansion joints and bond breaks should be provided to accommodate differential movement between rigid barriers and flexible wall faces. Cantilevered supports extending behind walls stabilize the barriers against overturning.

Fences

When there is sufficient space, the easiest and most cost-effective way to install fences above VERSA-LOK walls is to place them several feet behind walls. With sufficient fencepost depth and setback, the soil can provide a stable foundation. Separating fence posts from a wall also keeps wall movement from

affecting the fence. While a minimum post depth of thirty inches is suggested, the embedment and distance behind the wall needed to create a stable post foundation varies and depends on the soil and loading conditions.

When a fence is set back behind a wall, installers can dig or drill post holes after the wall is completed or they can install posts during wall construction. One option is to create postholes during wall construction by placing cylindrical tube forms at planned post locations and backfilling soil around them. After completing the wall, the tubes are filled with concrete and the fence posts set in the concrete.

The fence posts are installed several feet behind the edge of this wall, ensuring safety for students.

When there is not enough room to set fence posts behind walls, they can be installed within top wall units prior to backfilling behind the wall. Break off the backs of the top few units to create room for the post. Cut or core the cap units to neatly receive posts. The fence should be flexible enough to accommodate differential movement between the units and the fence.

Placing posts near the front of a wall decreases the fence's foundation support. To improve stability to the post, the concrete foundation should be enlarged, extended behind the wall and reinforced with steel rebar. The needed depth, extension length, and rebar placement will vary depending on conditions and loading.

Fencing and a gate added to Versa-Lok standard units provide both elegance and security to this common driveway.

Glossary

Backfill: Soil placed in front of and behind base course units. Also soil placed behind drainage aggregate. All backfill should be well compacted. Loose backfill will add pressure on walls, collect water, cause settlement, and will not anchor soil reinforcement materials properly. Backfill that is behind a wall containing soil reinforcement is often referred to as reinforced soil.

Base Course: The base course is the first layer of VERSA-LOK units placed on the leveling pad. Extra care should be taken to ensure that all base course units are level front to back, side to side, and with adjacent units. Unevenness in the base course becomes magnified throughout succeeding courses and cannot be easily corrected.

Bond: The arrangement or pattern of units from course to course. A unit that is centered over the joint created by the adjacent lower course units is placed on "1/2 bond." VERSA-LOK's unique pinning system permits variable-bond construction and allows units to fit close to each other while interlocking correctly. In general, VERSA-LOK units should be installed on 1/4 to 3/4 bond — where units overlap vertical joints of adjacent lower course units by at least four inches.

Compaction: Applying mechanical force to soils so they are no longer compressible. It is important to compact foundation and backfill soils to prevent future wall movement. Compaction is often accomplished using a hand tamper or a vibratory-plate compactor (available at most rental stores).

Course: A horizontal layer of retaining wall units.

Drain Pipe: Typically, a four-inch perforated pipe placed behind the wall at the base of the drainage aggregate. The drain pipe helps to direct large amounts of water from behind the wall to areas where it can accumulate safely away from the wall.

Drainage Aggregate: Clear, free-draining, angular gravel placed directly behind retaining wall units to expedite drainage. Drainage aggregate should not contain fine particles that could impede water flow.

Embedment: VERSA-LOK segmental retaining walls should have at least one-tenth of exposed wall height embedded below grade. For example, a four-foot wall should have approximately five inches of the base course buried below grade. Embedment provides enhanced wall stability and long-term protection for leveling pads. Embedment should be increased for special conditions such as slope at wall base, soft foundation soils, and shoreline applications.

Grade: The ground level, or ground elevation.

Gravity Wall: A retaining wall without soil reinforcement where unit weight alone provides resistance to earth pressures. Gravity walls are generally less than four feet in height and do not support slopes or other loads above the walls.

Impervious Fill: Backfill placed above and below the drainage aggregate. Impervious fill helps to prevent large amounts of water from running down behind the wall or getting to the leveling pad. Generally, compacted fine-grained soil is used as impervious fill.

Leveling Pad: The base on which a wall is constructed. Leveling pads consist of well-compacted crushed stone, gravel, or coarse sand. The most commonly used material for leveling pads is that which is used locally as road base aggregate.

Load: Weight or pressure placed on a retaining wall — usually from the back or top. Nearby slopes, driveways, buildings, and tiered walls all represent potential loads on retaining walls. Designs for retaining walls that support loads should be reviewed by a qualified, licensed professional engineer.

Saw Cuts: Saw cuts are made to modify VERSA-LOK units when smaller pieces are needed. Saw cutting creates a smooth, straight surface to meet cleanly with an adjacent unit. Saw cuts are generally made using a gas-powered cut-off saw equipped with a diamond blade – available at most rental stores.

Setback: The distance that each course is aligned behind the preceding (lower) course. Each course is set back 3/4 of an inch from the front of the course beneath it. This arrangement causes walls to cant back into the retained soil. Canted walls are structurally more stable than vertical walls because gravitational forces "pull" walls into retained soil.

Soil-Reinforced Wall: A retaining wall that incorporates horizontal layers of soil reinforcement material behind the wall. Soil reinforcement combines with soil to create structures that are strong and massive enough to support large loads. Soil-reinforced walls generally require a design by a qualified, licensed professional engineer (P.E.).

Soil Reinforcement: High-strength, polymer geosynthetic material, such as fabric or geogrid, that is buried in horizontal layers behind soil-reinforced retaining walls.

Split: Splits are made to modify VERSA-LOK units — most commonly to create corner units. Splitting creates attractive, textured surfaces identical in appearance to front faces of units. VERSA-LOK units can be easily split using a hammer and masonry chisel or mechanical splitter. This capability allows the user to create a wide variety of corners.

Split-face: The attractive, textured design on the front of non-weathered VERSA-LOK units. VERSA-LOK units are manufactured in pairs connected at their faces. They are then split apart, creating this appearance.

Tiered Walls: Two or more stacked walls with each upper wall set back from the underlying wall. Tiered walls can be attractive alternatives to single tall walls and can provide areas for plantings. To prevent an upper wall from placing a load on a lower wall, the upper wall must be built behind the lower wall a distance of at least twice the height of the lower wall. Tiered wall designs should be reviewed by a qualified, licensed, professional engineer.

VERSA-Lifter: The VERSA-Lifter speeds installation of VERSA-LOK retaining walls by making it easier to lift and place units — especially on the base course. The two prongs of the lifter are inserted into pinholes in the VERSA-LOK unit. The action of lifting the handle secures the lifter to the unit and makes for easy, balanced lifting and placement.

VERSA-LOK Adhesive: VERSA-LOK Adhesive is specially formulated to bond VERSA-LOK Cap Units to the top of walls. This adhesive remains flexible to accommodate the minor wall movement that may occur during freeze/thaw cycles.

VERSA-LOK Cap Units: VERSA-LOK cap units are used to attractively finish the top of VERSA-LOK retaining walls. Cap units come in two styles: A caps and B caps. Both cap styles are fourteen inches wide at the front, but A caps taper to twelve inches wide at the rear and B caps increase in width to sixteen inches at the rear. To cap straight walls, alternate A caps and B caps. Use A caps to finish walls with outside curves and B caps to finish walls with inside curves. Front faces of cap units may be placed flush, set back, or slightly overhanging (recommended) the face of the uppermost course of VERSA-LOK Standard units.

VERSA-LOK Retaining Wall Units: Premium retaining wall units offering easy installation, unmatched design flexibility, and unsurpassed durability. Units are installed without mortar or concrete footings. An unlimited variety of curves, corners, columns, and stairs can be constructed using only standard units. These environmentally safe, solid concrete units provide a lifetime of virtually maintenance free performance.

VERSA-LOK Weathered Retaining Wall Systems: VERSA-LOK Weathered units have rough, worn, rounded corners and their faces are jagged and rocky. Of course, VERSA-LOK Weathered units possess the same solid, pinned characteristics as all other VERSA-LOK Retaining Wall Systems.

VERSA-TUFF Pins: VERSA-TUFF nylon/fiberglass pins are used to interconnect VERSA-LOK retaining wall units and help provide consistent alignment. Pins are non-corrosive and will not rust or stain wall faces.

Chapter 1
For Homes and Yards
Creative Ideas for Home Landscaping

The staggered walls add scale to the landscape, making an expansive area appear more intimate and inviting.

This curved wall smoothes the transition from porch to lawn.

This clever homeowner decreased the mowing space by adding a raised planting space.

No back-aches from weeding with waist-level retaining walls.

Mixing mulch and stone lessens garden upkeep and adds to the visual appeal of the space.

Retaining walls with stairs and flowers built-in provide a lovely sunken patio, perfect for entertaining.

A curved stairway provides ground-level access to a raised patio with a commanding view.

Wide pillars, matching wall, and an iron fence add architectural interest to the landscape.

Varying heights in trees and plantings help punctuate a two-tiered raised bed.

This carefully crafted backyard has a jungle-gym
for the kids and a jungle of flowers for the adults.

Alternating the vegetation and terracing a slope cuts maintenance time and cost in the long run.

Dual gardening areas liven-up a sloped lawn.

A retaining wall creates a raised lawn where the family can step outside and enjoy both grass and view.

Retaining walls unify the home, a patio area, and a raised pavilion in this impressive backyard environment.

A round of gardens and columns boosts this sunroom patio.

A semi-circle of wall, punctuated by columns and multi-levels, eases the transition from deck to lawn.

The addition of a patio to this home led to a sunroom as well. Homeowners can enjoy the sun outdoors or indoors.

Retaining wall product works as a perfect surround for an in-ground spa, creating a custom look for the fiberglass unit.

One can relax in the shade without lawn care hassles, thanks to the patio and mulch design of this yard.

A steep yard can be easily conquered with retaining walls. The walls punctuate the slope and offer many landscaping possibilities.

The decorative rock adds a unique quality to this sunken patio.

A Versa-Lok wall was used to reclaim a portion of this sloping yard and create a
raised paver patio.

For a bolder design, use wide stairs and crisp corners.

A flower-lined sunken patio adds an aura of privacy and intimacy to a party.

Chapter 2
The Walls, in Detail
Looking Closely at the Materials and Magic of Retaining Walls

A retaining wall bisects a hillside, creating a property barrier as well as a more manageable landscape.

The sheer immensity of this wall is softened with a random color palate, mixed block sizes, and a gentle wave configuration.

Mixed colors and heights add interest to a long stretch of wall bordering a ballpark outfield.

Historical landmark uses retaining wall to complement surroundings.

Ready for planting, a multifaceted wall stands in contrast to a sand-filled environment.

Set-backs on each course of block create a sloped wall.

This downward sloping wall frames a shady walkway.

An example of how a short wall can add many pleasing aspects to a small yard.

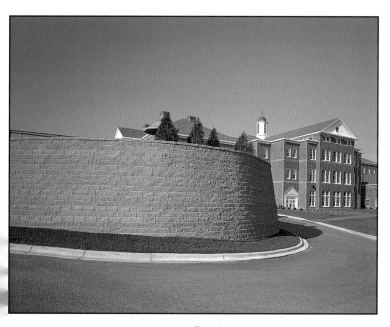

Retaining walls may add visual interest while holding hillsides at bay.

Chapter 3
Gardening
Using Flowers and Plants for a Beautifully Landscaped Home

A retaining wall created the opportunity for a developer to offer a level lawn area while installing a sidewalk. The homeowner took advantage of the amenity by creating a stretch of flowerbed.

The use of contrasting flowers and various levels is an opportunity to be creative and have the best lawn in the neighborhood.

A stretch of stone, underlined with landscaping fabric, provides maintenance-free drainage between a parking area and the garden.

Bright flowers, white capstones, and warm block combine for eye appeal in this garden nook.

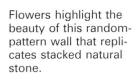

Flowers highlight the beauty of this random-pattern wall that replicates stacked natural stone.

This elegant curve of stand-alone wall provides a planter for a splash of colorful flowers, as well as convenient seating.

Just like the statue, one can sit forever in the garden created by this staggered retaining wall design.

Stair-step drops in height trace the slope of this two-tiered wall.

Low-maintenance gardens
of mulch and bushes
appear more sophisticated.

Lawn care is no longer
a hassle for this
homeowner. Stone and
shrubbery reduce the
need for mowing. Knee
walls around the patio
area provide seating.

Columns and fencing cap a two-tiered wall encircling the backyard of this property. The wall permitted the installation of a pool on a sloped lot, and the fence provides required protection around the water.

A sloping retaining wall and small garden add to the appeal of a stroll down this sidewalk.

Multi-colored paver stones are beautifully framed by tan walls and capstones.

Adding a mantle to a retaining wall is an easy step and adds many decorating possibilities.

Birds can enjoy this garden as well.

Incorporating many shapes, flowers, and plants will create a distinctive display.

The home front is extended with a curved landing and raised garden.

Make the most out of a small backyard by installing adjacent retaining walls for gardening space.
Adding a statue can ensure safety and good harvests.

A flowerpot highlights the crisp angles of this free-standing wall.

Chapter 4
Life with a Retaining Wall
Demonstrating the Luxury and Utility

Lights installed throughout a free-standing wall create a romantic ambiance for this patio.

A circular planter brightens an otherwise normal entryway and coordinates with the paver drive.

A retaining wall was used to build a raised paver patio level with the walkout lower level family room.

Parallel pillars and rounded steps make an elegant back porch entrance.

Retaining walls, stairs, and a circular landing combine to create a beautiful walkway through a sloping garden.

One can spend countless hours enjoying the sun and relaxing on this circular paver patio.

A half-wall provides built-in seating for this outdoor kitchen/patio area.

The only thing missing is the kitchen sink in this outdoor kitchen.

A retaining wall serves as home to kitchen appliances, and was capped with durable countertop material.

An outdoor environment is made all the more useful with on-site cooking and seating. The wall, which forms a backdrop, is also useful for seating, or as a countertop for cooking and serving food.

A gas-fired grill and cooler share real estate in a retaining wall.

This couple is enjoying a moment together, nestled in a patio environment created with multi-leveled retaining walls.

A secluded spot for relaxation on a crisp autumn day.

This squared-off patio with a raised garden area and planter is the perfect spot to plant oneself.

Pillars and flowers help form a secret hideaway.

A knee-wall defines this
patio space and provides
the perfect environment for
garden enthusiasts.

Juxtaposing right angles and half circles provides room in the middle for entertainment and levels for decorating.

A knee wall encircles an intimate patio, and cuts down on the need for furnishings.

A fire-pit and knee-wall await evening gatherings.

The balcony and lower level of this home are accentuated by the circular staircase landing and two-tiered retaining wall.

Walls define the "rooms" in this outdoor environment

A stairway descends to a cozy patio area.

A circular retaining wall makes a display area for this rock.

This family strategically planned their patio: the grill is close to the kitchen, the fire pit is away from the house, and the furniture is off to the side to ensure mingling in the middle.

A gradual staircase leads up to a stone-speckled lawn.

This gorgeous layout defines luxury with a fire pit and hot tub.

The curved lines and brick pattern add visual stimulation to the fire-pit experience.

The trapezoid shape of this patio was constructed by cutting and splitting Versa-Lok Standard units.

Together, a half wall and railing add safety barriers
and seating for a patio.

A circular retaining wall above ground creates a fire pit. Many stories are waiting to be told and marshmallows to be melted. Logs are stacked close by to keep the fire going.

The use of columns and gray stones creates a bold but elegant patio area.

Staggered heights create a
border around a circular patio.

This sunken patio is framed with gorgeous flowers and creates a picture of privacy.

Outdoor dining is finally safe from wind blown tables! The base of this table is made of solid Versa-Lok units and is capped with a slab of natural stone.

This narrow staircase leads to a wide patio area, just waiting for a visitor. The column is capped with a lamppost to guide one down the stairs or onto the lawn.

A small barrier is created around this patio area, ensuring some privacy.

This freestanding wall distinguishes the patio from the garden and creates a clever hiding space.

Chapter 5
Retaining Walls with Water
From Poolside to Lakeside

Fences, stacked levels, and descending stairs accent this gorgeous pool environment.

A half wall helps restrict critter access to the pool area.

A wall's course continues as a pool border, its course broken only by inset stairs.

Parents can watch their kids swim from the comfort of their own patios thanks
to the pool area created by a high retaining wall.

Miles of smiles and slides await the owners of this pool. The structure used to build the slide is a perfect storage location
for toys, towels, and other pool paraphernalia.

An expanse of wall is softened and brightened with a fall of trailing plantings.

Those who don't like to be splashed while sunning poolside can find a dry haven on the second level of this pool patio. A mighty wave of water would be needed to soak the sunbathers.

A large deck for dancing and pool for swimming make this a great summer hangout. The area between the two tiers provides a decorative space and support.

This patio is ideal for day-dreaming with views of both the river below and the pool, all made possible by the walls that buck up this patio from beyond.

This in-ground pool was built by adding retaining walls in return for extra support and drainage.

Cool summers and fun times await this family.

Chapter 6
Step by Step
Using Retaining Walls to Create Stairways

Retaining wall and steps were a necessity for making this backyard negotiable.

Three tiers for gardening launch a staircase to the back deck.

A bird's-eye view accentuates the impact that a retaining wall can create on the landscape.

Landings and angled climbs help ease the transition up for the pedestrian. Large platforms afford places where pedestrians can catch their breath and appreciate the view.

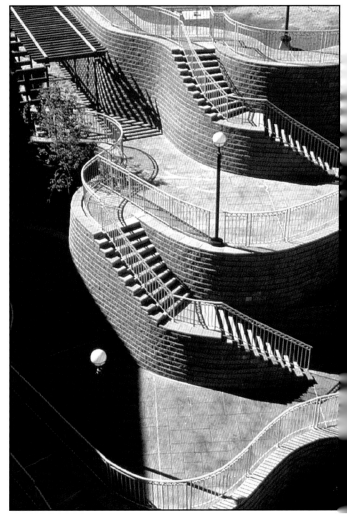

Walls not only conquer this hillside, but they also provide handrails and resting spots along a stretch of stairs.

Sand-colored block looks at home amidst the fruits of a green thumb.

Rich, stone-toned walls create a warm underline for this towering home.

An elegant staircase intersects three tiers of terrace.

This split stairway adds fun to the front yard and conquers a steep hill.

Wide steps and creative landscaping ease the transition from porch to yard.

An aerial view illustrates how retaining walls and a staircase intersect.

Properly constructed, this staircase and walls should stand up for centuries.

Walls flare at either side of the stair landing, providing a welcome gesture as one approaches.

Lighting enhances the appearance of a retaining wall, and adds a measure of safety when using the stairs.

A sense of grandeur is instilled with wide stairs and iron gates.

Stacked tiers, platforms, and dark stone are handsome amenities to the landscape.

This staircase was etched into the side of the hill to create picnic and play areas. This concept can be used for parks or schools.

Stairs, walls, and pillars coordinate to add to the beauty of this gorgeous home.

Mixed pavers create a quilt-like appeal within the borders of warm-toned walls.

Handrails are required safety elements for most public staircases.

Stairs make the lawn accessible from two directions below the home.

A staircase carved into this hill provides a direct route from the driveway to the pool.

The tall sides of this staircase serve as guides down the steep descent.

Spread out, staircases can make the climb seem less daunting. Flowerpots placed on the columns trick the eye into believing it's a diagonal ascent.

The different textures of the columns and the stairs create a calming contrast.

The side gardens make this staircase seem like part of the hill.

A twisting staircase is created by alternating the height of the tiered walls and the direction of the stairs.

Bright flowers and creative lawn ornaments make this staircase fun to climb.

Chapter 7
Having Fun with Retaining Walls
Recreational and Public Uses

Strong walls can facilitate a comfortable relationship between shore and bodies of water.

It's a homerun, if the ball gets past this retaining wall.

Retaining walls are common features in the manmade environment of a golf course.

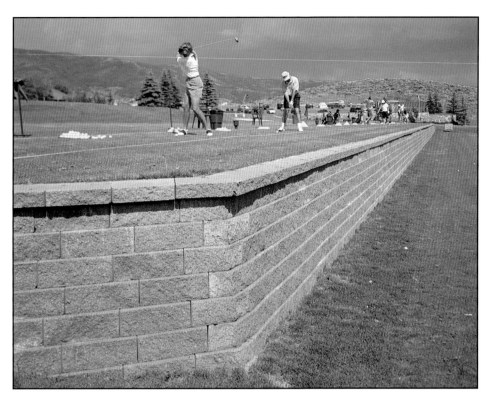

A retaining wall creates the launching point for driving practice.

Golfers also pass conveniently from one fairway to the next between these graceful retaining walls.

The design of this multi-leveled terrace created access for a hillside.

A fire pit, and firewood ready for chopping, found a home on this raised patio created by a sturdy retaining wall.

The residents of this community used a raised island garden to attractively identify their development.

Walls are the solution for a site too steep for vehicular access.

The staggered angles can make reversing a large truck much easier in this loading area.

Slightly stepping up, this wall adds some dimension to its appearance.

Bold columns and wide stairs add elegance to this lovely home.

With proper drainage and design, one can add a tranquil water fountain to any landscape.

This homeowner used freestanding planters, tiered walls, staircases, and a handicapped ramp to make his home attractive and easily accessible.

After a long day on the bay, one can easily climb the slope from the dock to the house thanks to a built-in staircase and terraced retaining walls.

This unique divider was created by adding split-rail fencing between columns. The wood joints can decrease the costs of a retaining wall or the columns add a stylish twist to a boring fence.